# MAPS
# OF OLD LONDON

I. WYNGAERDE (IN THREE SECTIONS)
II. AGAS
III. SECTION OF AGAS
IV. HOEFNAGEL
V. NORDEN LONDON
VI. NORDEN WESTMINSTER
VII. FAITHORNE
VIII. OGILBY
IX. ROCQUE

LONDON
ADAM AND CHARLES BLACK
1908

## EDITOR'S NOTE

An atlas of Old London maps, showing the growth of the City throughout successive centuries, is now issued for the first time. Up to a recent date the maps here represented had not been reproduced in any form, and the originals were beyond the reach of all but the few. The London Topographical Society has done admirable work in hunting out and publishing most of them; but these reproductions are, as nearly as possible, facsimiles of the originals as regards size, as well as everything else. It is not every one who can afford to belong to the society, or who wishes to handle the maps in large sheets. In the present form they are brought within such handy compass that they will form a useful reference-book even to those who already own the large-scale ones, and, to the many who do not, they will be invaluable.

The maps here given are the best examples of those extant, and are chosen as each being representative of a special period. All but one have appeared in the volumes of Sir Walter Besant's great and exhaustive "Survey of London," for which they were prepared, and the publishers believe that in offering them separately from the books in this handy form they are consulting the interests of a very large number of readers.

The exception above noted is the map known as Faithorne's, showing London as it was before the Great Fire; this is added for purposes of comparison with that of Ogilby, which shows London rebuilt afterwards. Besides the maps properly so called, there are some smaller views of parts of London, all of which are included in the Survey.

The atlas does not presume in any way to be exhaustive, but is representative of the different periods through which London passed, and shows most strikingly the development of the City.

I must acknowledge the valuable assistance I have received from Mr. George Clinch, F.G.S., in the many difficulties which arose in the course of its preparation.

G. E. MITTON.

# PANORAMA OF LONDON

## By ANTONY VAN DEN WYNGAERDE

**Description.**—This is the earliest representation of London that has come down to our time. Accurately speaking, it is not a map, but a picture; but as many of the old maps are more or less in the same category, we need not exclude it on that account. Such topographical drawings are apt to be misleading, owing to the immense difficulties of perspective — witness the wretched samples hawked about the pavements at the present time. But, considering the difficulties, this map of Wyngaerde's is wonderfully accurate, and it has the advantage of being full of architectural details which no true map could give.

**Designer.**—Of Wyngaerde himself little is known. He is supposed to have been a Fleming, and may have come to England in the train of Philip II. of Spain. He is known to have made other topographical drawings. The date of the one here reproduced cannot be fixed with perfect certainty, but must have been between 1543 and 1550.

**Original.**—The original is in the Sutherland Collection at the Bodleian Library, Oxford, and it measures 10 feet by 17 inches, and is in seven sheets. A tracing of it, made by N. Whittock, can be seen in the Crace Collection, Prints Department, British Museum, or in the Guildhall Library.

The present reproduction is from that made by the London Topographical Society, which photographed the original.

It is reduced, and is here placed in three sections, which overlap for convenience in handling.

## I.

**Details.**—If we examine the first section, which is that to the extreme west, we see the Abbey, very much as it is at present, with the exception of Wren's western towers. On the site of the present Houses of Parliament is the King's Palace at Westminster. It is impossible here to treat this in detail, for if that were attempted for all the buildings in this atlas, space would fail. A concise account of Westminster may be found in the book of that name in the *Fascination of London* Series. The chief point to note in the palace is St. Stephen's Chapel, of which the crypt now alone remains. About fifteen or twenty years previous to the date of this map King Henry VIII. had claimed Whitehall from Wolsey, and transferred himself to it from the old palace, which was growing ruinous.

Across the river opposite to Westminster is Lambeth, standing in a grove of trees.

Beyond Westminster westward all is open ground, in the midst of which we see St. James's Hospital, where is now St. James's Palace. Though still marked "Hospital," it had already been annexed by the King. Where is now Trafalgar Square we are shown in the map the King's Mews, built by Henry VIII. for his hawks. Charing Cross is marked by the cross put up in memory of Queen Eleanor. Along the river banks is a fringe of fine houses and foliage. We may pick out one or two of these princely buildings—namely, Durham House, Savoy Palace, and Somerset House (see *The Strand* in the above series). The church of St. Clement Danes is only separated from the open country by a single row of houses.

On the west side of the Fleet River is Bridewell, built by Henry VIII. in 1522 for the entertainment of the Emperor Charles V. Here, in 1529, Henry and Katherine stayed while the legality of their marriage was being disputed in Blackfriars across the Fleet. Then we come to Old St. Paul's, still carrying its tall spire, destined so soon to topple down. Between it and the river is one of the most famous of the old strongholds, Baynard's Castle. On the extreme right of the map is the port of Queenhithe, which can be seen to-day by any wanderer in the City.

II.

Turning the page, we see the old City as it was before the Fire, made up of gable-ended wooden houses with overhanging stories, crowded close together, and diversified by the numerous pinnacles and spires of the City churches, many of which were never rebuilt. The embattled line of the wall hems the City in on the north, and Cheapside cuts it laterally in a broad highway. Almost in the centre of the picture is the Guildhall. The interest reaches its culmination in the spectacle of Old London Bridge, with its irregular houses, its archways, and its chapel. Note that the engraver has not omitted to indicate the decaying heads on poles, a succession of which adorned the bridge throughout the centuries (see *The Thames* in above series).

On the south side of the water is St. Mary Overies (see *Mediæval London*, vol. ii., p. 29,). It has as neighbours Winchester and Rochester Houses, the residences of the respective Bishops of those sees; while the proud cupolas of Suffolk House— built *circa* 1516, and later used as the Mint—are clearly shown. The houses running from it up to the foreground of the picture are beautifully delineated, and may be taken as models of Elizabethan architecture; while the man with the harp and the horseman are quite clearly enough drawn to show their period by the style of their dress. From some point behind here must Wyngaerde have made his survey, as it is manifestly impossible it could have been done from Suffolk House, as stated by one authority.

### III.

There are three objects so striking in this picture that attention is at once claimed by them to the exclusion of all else—the Abbey of Bermondsey, the Tower of London, and Greenwich Palace. In Bermondsey two Queens died— Katherine, consort of Henry V., and Elizabeth, consort of Edward IV. Only a year or two before this map was made had the grand old Abbey been surrendered to the King (for a full account see *Mediæval London*, vol. ii., p. 288).

The Tower, taken as a whole, is very much as we still know it; it is one of the oldest remaining relics of the past. Note the gruesome place of execution near by, and the guns and primitive cranes at work upon the wharf. Just beyond it eastward rise the fretted pinnacles of St. Katherine's by the Tower, on the spot now covered by St. Katherine's Docks.

Stepney Church stands far away on the horizon, cut off from the City by an ocean of green fields.

Returning to the south side, we see Say's Court, Deptford, between Bermondsey and Greenwich. This was for long the home of John Evelyn, and was ruinously treated by Peter the Great, who tenanted it during his memorable stay in this country in 1698. (For Greenwich Palace or Placentia, see *London in the Time of the Tudors*.)

# CIVITAS LONDINUM

Description.—This is the earliest map of London known to be in existence, for though Wyngaerde's survey preceded it in date, as we have seen, that is a panorama and not a map proper. The present map, which is known as that of Ralph Agas, itself has a good deal more of the panoramic nature than would be allowed in a modern one, and is on that account all the more interesting. The first to connect Agas's name with this map was Vertue (1648-1756), and he stated its date to be 1560; but, as will be seen in the description of the next plate, Vertue's claims to strict veracity have now been shaken, therefore his testimony must be accepted with caution.

Designer.—Ralph Agas, land surveyor and engraver, died in 1621, and he is described in the register as "an aged." Of course, it is possible that Agas lived to the age of eighty-five or over, in which case he might not have been too young to execute this work in 1560, and he himself says, in a document dated 1606, which has been preserved, that he had been in work as a surveyor for upwards of forty years. There are two branches into which the enquiry now resolves itself. First, did Agas really make the map? And, second, if he did, at what date did he make it? There is no conclusive evidence on either hand. There is a survey of Oxford, similar in character, signed by him, and though this is not dated, it is known to have been completed in 1578, and published ten years later. On the original copy of this, which is at the Bodleian, there are the following lines:

> "Neare tenn yeares paste the author made a doubt
> Whether to print or lay this worke aside
> Untill he firste had London plotted out
> Which still he craves, although he be denied
> He thinkes the Citie now in hiest pride,
> And would make showe how it was beste beseene
> The thirtieth yeare of our moste noble queene."

Original.—The two earliest known copies of the Agas map, which was first engraved on wood, are both of the same issue; one is at the Pepysian Library, Magdalen College, Oxford, and the other at the Guildhall. Edward J. Francis made a careful reproduction of that at the Guildhall in 1874, and it is from that our present plate is taken. It is, of course, reduced, for the original is 6 feet and ½ inch long, by 2 feet 4½ inches wide. The notes attached to this issue are by W. H. Overall, F.S.A., one of the leading authorities on the question. He doubts Agas's connection with the map, but thinks if he were the originator it could not have been done before 1591. The arms in the corner on the two oldest extant maps are those of James I., but as the arms on the royal barge in the river are those of Elizabeth, it has been conjectured that the maps are themselves copies of a later edition, wherein the arms were altered in conformity with conventional opinion. The chief points which give data from internal evidence are as follows: St. Paul's Cathedral is bereft of its spire. This was struck by lightning in 1561, so the map must be subsequent to that date. The Royal Exchange is apparently built. This was opened in 1570. Northumberland House, built about 1605, has not been begun. We may take it, therefore, generally that the original map, which was engraved on wooden blocks, was made some time in the latter half of Elizabeth's reign, and it is probable that it was done by Agas.

Details.—The map abounds in interesting detail.

Beginning in the extreme left-hand lower corner, we see St. Margaret's Church, St. Stephen's Chapel, and Westminster Hall. In the river are swans of monstrous size. King Street, now merged in Whitehall, is very clearly shown, also the two heavy gates barring the way. The most northern of these, designed by Holbein, was called after him, and stood until the middle of the eighteenth century. North of it, on the west, is the tilting-ground; and stags browse in St. James's Park. Between the gates, on the east, are the Privy Gardens, overlooked by the Palace of Whitehall—most unpalatial in appearance.

Piccadilly is "the Waye to Redinge," and Oxford Street "the Waye to Uxbridge." Near Whitcomb Lane and the Haymarket women are spreading clothes in the fields to dry, while cows as large as houses graze around. St. Martin's Lane leads up to St. Giles, more particularly dealt with in the description of the next plate. The irregular buildings of St. Mary Rouncevall, a religious house, had not yet been taken down to make way for Northumberland House, itself to be replaced by Northumberland Avenue. The houses of great nobles, with their magnificent gardens stretching down to the waterside, are still in evidence. North of the well-laid-out Covent Garden, owned by the Dean and Chapter of Westminster, are nothing but trees and fields. Passing on quickly down the Strand, we find Temple Bar blocking the way to the City. This is the old Temple Bar, replaced after the Great Fire by the one much more familiar to us, which stood until 1878. A very fine illustration of the old one is given in Sir Walter Besant's *London in the Time of the Tudors*, p. 245. This book should certainly be studied by anyone desirous of understanding the map. From Temple Bar past the back of St. Clement's Church runs a broad road roughly corresponding with our new Kingsway. Further eastward the Fleet River still flows strongly down from its northern heights, crossed by many bridges, and just where it joins the Thames is Bridewell Prison. Further along, on the other side, is Baynard's Castle, and in front of it, in the river, the Queen's barge, with the royal arms of Elizabeth in the centre. Some way back from Baynard's Castle a bridge crosses a street, and is marked " The Wardrop." This was in very truth the wardrobe or repository of the royal clothes! Drawing a line northward for some way, we come to Smithfield, where tilting is represented as in animated progress. Not far northward is St. John's, Clerkenwell, and its neighbouring nunnery; to the west is the Charterhouse. Turning south again, past St. Bartholomew's Church, we see the building of Christ's Hospital, founded by Edward VI. This, it may be noted, is one of the buildings erected since Wyngaerde's time. Then we come to St. Paul's, shorn of its spire, with St. Gregory's Church, quite recognizable, in front of it. There were continual edicts against building in the Tudor and Stuart reigns, for it was feared London would grow out of hand; but, in spite of this, houses have enormously increased since Wyngaerde made his survey. The battlemented wall still encloses the City, but hamlets have sprung up outside, notably at Cripplegate.

But within the wall there are still some fine gardens and open spaces, one of which remains to this day in Finsbury Circus. Many roads meet in the heart of London, where now the Bank, Mansion House, and Royal Exchange stare across at each other. It is difficult to make out from the medley of buildings in the map if Gresham's first Royal Exchange is there or not, but it seems to be so. This was opened in 1570 by the Queen in person. St. Christopher le Stock's square tower may be seen on the ground now absorbed by the Bank of England.

Crossing over now to the Surrey side, we see conspicuously the two round pens for bull- and bear-baiting respectively. There are many pleasure-gardens, for the Surrey side was for long the recreation-ground of the Londoner. On the river there are innumerable wherries, and below the bridge at Billingsgate many ships cluster; one has even managed to get above the bridge. Off the Steelyard and at the Tower are men and horses in the water. This is a most interesting point. In those at the Tower it may be clearly seen that the man is filling the water-casks on the animals' backs with a ladle. This gives a glimpse into the discomforts endured by our ancestors before water-pipes were laid on as a matter of course to all houses. In the eighteenth-century reproductions of this map, oddly enough, in one instance this detail has disappeared, and in the other it is turned into a man driving cows into the water with a whip; thus doing away with all its significance. Far to the north in Spitalfields men are practising archery; while Aldgate, for long the home of Geoffrey Chaucer, is conspicuous a little north of the Tower.

As became a man living in days of the Reformation, Agas does not point out the religious houses then falling into decay or occupied by laymen, yet what a number of them must have been still in existence! Standing on the White Tower, and looking north and to the right hand, there must have been visible outside the wall

St. Katherine's by the Tower, Eastminster, and the Sorores Minores, whose name still remains in the Minories, here marked. Within the City was Holy Trinity, close to Aldgate—of this a couple of most rare and interesting plans and a full account may be found in *Mediæval London*, vol. ii.—and not far off was St. Helen's Nunnery; also Crutched Friars, Austin Friars, Grey Friars, and, in the extreme west, near the Fleet, Blackfriars. Of these and many others full accounts may be found in the volume indicated above.

# THE PARISH OF ST. GILES IN THE FIELDS

**Description.**—This plate, on being compared with the preceding one, shows a strong general resemblance, with a considerable difference in detail. Also, below are two churches, one of which is marked, "Present St. Giles's Church, built anno 1734," which shows that the map was made not earlier than that date. It is, in fact, a part of one of a set of eighteenth-century maps based on that of Agas, and not only differing from it in detail, but also differing slightly one from another. Some of these are unsigned, and some are signed "G. Vertue," and were specifically claimed by Vertue as having been made by him, and based upon Agas's map of 1560. Recently, however, doubts have been raised as to Vertue's share in the transaction, and it is now very commonly believed that he did no more than procure some maps, engraved on pewter and made in Holland, based on that of Agas. These he altered a little in detail, and then claimed as his own work. The original pewter plates are in possession of the Society of Antiquaries, Burlington House. The present example differs in some small particulars from these. Copies of the maps are not rare, and can be seen at the British Museum and elsewhere.

**Details.**—The bit of London here represented is of exceptional interest. It shows the corner of Tottenham Court Road when High Street and Broad Street, St. Giles, were the main highway, long before the cutting through of New Oxford Street. It shows, further, the descent of Holborn into the valley of the Fleet, the "heavy hill" along which criminals were brought from Newgate to the place of execution. It shows the site where the gallows stood for some time, about 1413, before being definitely set up at Tyburn. Close to this was the Bowl tavern, where the condemned man was allowed his last draft of ale. The most interesting old hospital for lepers is clearly shown. (See "Holborn," *Fascination of London* Series.)

# "LONDINUM FERACISSIMI ANGLIÆ REGNI METROPOLIS"

### By HOEFNAGEL

Description.—This map seems at first sight to be much less interesting than those which have preceded it, but that is due chiefly to its small size. The probable date is 1572, and even if otherwise unknown, it might have been judged approximately by the costumes of the figures in the foreground. It must have been contemporary with, or even earlier than, Agas, with whose work it is interesting to compare it. This map was made by Hoefnagel, and is taken from Braun and Hogenburg's work, *Civitates Orbis Terrarum*, in which Braun wrote the text, while Hogenburg and Hoefnagel engraved the maps. In the left-hand top corner are the arms of Elizabeth, and in the right-hand corner those of the City. In the later editions the delicately drawn figures in the foreground are omitted. In his notes on Old London Maps in the Proceedings of the Society of Antiquaries, vol. vi., Mr. W. H. Overall says it cannot be supposed that all the cities of the world engraved in Braun and Hogenburg's work were freshly surveyed for the purpose; and there are several points—such, for instance, as the inclusion of the steeple of St. Paul's, destroyed in 1561—which point to the fact that this version was probably taken from existing surveys. The original is 19 inches by 12¾ inches. The bull- and bear-baiting pits on the Surrey side are quite conspicuous, and so is the royal barge, in very much the same position in the river as it is in Agas's map. Here is a detailed account of it in Sir Walter Besant's own words:

Details.—"This is in some respects more exact than the better-known map attributed to Agas. The streets, gardens, and fields are laid down with greater precision, and there is no serious attempt to combine, as Agas does, a picture or a panorama with a map. At the same time, the surveyor has been unable to resist the fashion of his time to consider the map as laid down from a bird's-eye view, so that he thinks it necessary to give something of elevation.

"I will take that part of the map which lies outside the walls. The precinct of St. Katherine stands beside the Tower, with its chapel, court, and gardens; there are a few houses near it, apparently farmhouses. The convent of Eastminster had entirely vanished. Nothing indicates the site of the nunnery in the Minories, yet there were ruins of these buildings standing here till the end of the eighteenth century. Outside Bishopsgate houses extended past St. Mary's Spital, some of whose buildings were still apparently standing. On the west side St. Mary of Bethlehem stood, exactly on the site of Liverpool Street Station, but not covering nearly so large an area; it appears to have occupied a single court, and was probably what we should now consider a very pretty little cottage, like St. Edmund's Hall, Oxford.

"Outside Cripplegate the houses begin again, leaving between the Lower Moorfields dotted with ponds; there are houses lining the road outside Aldersgate. The courts are still standing of St. Bartholomew's Priory, Charterhouse, St. John's Priory, and the Clerkenwell nunnery; Smithfield is surrounded with houses; Bridewell, with its two square courts, stands upon the river bank; Fleet Street is irregular in shape, the houses being nowhere in line; the courts of Whitefriars are still remaining. The Strand has all its great houses facing the river; their backs open upon a broad street, with a line of mean houses on the north side. On the south of the river there is a line of houses on the High Street, a line of houses along the river bank on either side, and another one running near Bermondsey Abbey.

"Within the walls we observe that some of the religious houses have quite disappeared—Crutched Friars, for instance. There is a vacant space, which is probably one of the courts of St. Helen's. The Priory of the Holy Trinity preserves its courts, but there is no sign of the church. There are still visible the courts and gardens of Austin Friars. There is still the great court of the Grey Friars, but the buildings of Blackfriars seem to have vanished entirely" (*London in the Time of the Tudors*, p. 185).

# NORDEN'S MAPS OF LONDON AND WESTMINSTER

**Designer.**—Being on a very small scale, these maps are not so attractive as some that have been already discussed. John Norden, the designer, was born about 1548, and seems to have had from the first an extraordinary gift of delicate penmanship, which he turned to much account in map-making. He projected a whole "Speculum Britanniæ," but during his lifetime only managed to publish books on two counties—namely, Middlesex and Hertfordshire. He left behind him the results of his labours on many other counties in manuscript, and these have since been published. Norden was appointed Surveyor of His Majesty's Woods in 1609. The engraving of the Middlesex maps was done by Peter Van den Keere.

**Originals.**—The reproductions are taken from those which appear in Norden's *Middlesex*, dated 1593. Each map is 9½ inches by 6¾ inches. The wonderful delicacy of Norden's work makes these maps peculiarly appreciated by students of London cartography.

# FAITHORNE AND NEWCOURT

**Description.**—This map generally goes by the name of Faithorne, the engraver, but in reality the credit is due quite as much to Richard Newcourt the elder (d. 1679), who was the draughtsman. It is selected for a place here because, the date being 1658, it shows the City as it was before the Fire, and therefore forms a supplement to the map of Ogilby which follows, and shows the City as it was when rebuilt after the Fire.

**Engraver.**—William Faithorne the elder was born in 1616, and was an engraver and portrait painter. He engraved numerous portraits, book-plates, maps, and title-pages. Among his works are two large maps, entitled "Cities of London and Westminster," and of "Virginia and Maryland."

**Original.**—The only two copies of the original issue known to be extant are in the Print Rooms, British Museum, and in the Bibliothèque Nationale of Paris. The map here given is taken from a sheet of that in the British Museum, and is on the same scale.

**Details.**—It will be noticed that the sheet chosen for inclusion in this atlas shows very nearly the same area as the map of Ogilby which follows, but does not go quite so far eastward as the Tower. The City wall is clearly shown along the north side of the City, and the bastion near Cripplegate stands out; the town ditch can be traced just beyond this corner running southward. It was the curious and apparently meaningless angle that the wall makes here which led Sir Walter Besant to suggest that it may have been designed to exclude the ancient Roman amphitheatre, of which the site is now lost (see *Early London*, p. 85). The Fleet River is shown still open and crossed by bridges, of which there are no fewer than five from Holborn to the mouth. That at Fleet Street shows, indeed, a continuous line of houses. St. Paul's is very clearly delineated. The figures within the City refer to the old churches, of which a list is given below. Notice the gable roofs, still the chief style of domestic architecture. The lines of the streets in the heart of the City remain wonderfully the same to our own day. Outside the walls the City is stretching out great arms into the country. There is one such arm made by the continuous houses fringing Bishopsgate Street as far as the extreme northern limit of the map. Then there is a gap between this and Moorgate Street, including all the ground known at Moorfields and Finsbury. A few scattered houses and some cultivated fields cover this space, and in one corner is "Bedlame."

A mass of houses lies westward, running on to the Charter House, northward of which are open fields, and so to "Clarkin Well."

THE SEVERALL CHVRCHES WITHIN THE WALLES OF LONDON DISTINGUISHED BY SEVERALL FIGURES, BY WHICH ALLSOE THE EYE MAY PARTLY BE GUIDED TO THE EMINENT STREETS IN OR NEERE WHICH THEY STAND, WHICH COULD NOT WELL BE OTHERWISE DEMONSTRATED, IN REGARD OF THE SMALL SCALE BY WHICH THIS MAPP IS DESCRIBED.

1. Albans in Woodstreet
2. Alhallows Barkin nere Tower hill
3. Alhallows in Bread street
4. Alhallows ye Greate in Thamas
5. Alhallows the Lesse streete
6. Alhallows in Hony lane nere Chepside
7. Alhallows in Lumber street
8. Alhallows Stayninge nere Fanshawes street
9. Alhallows in ye Wall nere Moorefeilds
10. Alphage by ye Wall nere Cripplegate
11. Andrew Hubard by Philpot lan
12. Andrew Vndershaft
13. Andrew in ye Wardrop aboue Pudle wharfe
14. Ann at Alders gate
15. Ann in Black friers
16. Antholins in Watling streete
17. Austins nere Paules church
18. Bartholomew by ye Exchange
19. Bennet Finch
20. Bennet Grace church neer Gracious streete
21. Bennet at Paules wharfe
22. Bennet Sherehogg nere Bucklersberry
23. Bottolph at Billings-gate
24. Christs Church by Newgate streete
25. Christophers in Thredneedle streete
26. Clements in East chepe
27. Dennis back Church nere Eashastreete
28. Dunstanes in ye East nere Tower street
29. Edmonds in Lumber streete
30. Ethelborough in Bishops gate street
31. Faith under Paules
32. Foster in Foster lane nere Chepside
65. French Church in Third needle street
33. Gabriell in Fanshawes streete
34. Georges in Bottolph lane
35. Gregories by Paules
36. Hellins nere Bishops gate
37. James Duke place nere Aldgat
38. James Garlick hill by Bow lane
39. Iohn Baptist nere Dow gate street
40. Iohn Euangelist nere Friday street
41. Iohn Zachary nere Foster lane
42. Katherin Coleman nere Fanshawe stret
43. Katherin Cree church nere Aldgate
44. Lawrence Iury nere Guild hall
45. Lawrence Poultney nere Eastchepe
46. Leonarde in East-chepe
47. Leonarde in Foster lane
48. Magnus by the Bridge
49. Margrett in Loth bury
50. Margrett Moses next Friday street
51. Margrett in new Fishstreete
52. Margrett in Rood lane
53. Mary Abchurch Lane
54. Mary Alderman berry
55. Mary Aldermary nere Watling streete
56. Mary le Bow in Chepside
57. Mary Bothaw in Cannon streete
58. Mary Cole church in Chepside
59. Mary Hill aboue Billings gate
60. Mary Mounthaw aboue Broken warfe
61. Mary Somersett nere Broken wharfe
62. Mary Staynings nere Alders gate
63. Mary Woollchurch nere ye Stocks
64. Mary Wooll noth in Lumber streete
66. Martins Iremonger lane nere Chepside
67. Martins within Ludgate
68. Martins Orgars nere Eastcheape
69. Martins Outwitch next Bishopsgate stret
70. Martins Vintree neere ye 3 Cranes
71. Mathews in Friday Street
72. Maudlin milke strct neere Chepside
73. Maudlins in Old Fishstreete
74. Michaell Bashaw behind Guildhall
75. Michaell in Cornhill
76. Michaell Crooked Lane nere N Fish'trete
77. Michaell att Quene Hith
78. Michaell ye Querne vper end of Chepside
79. Michaell Royall att Colledge Hill
80. Michaell in Woodstreet nere Chepside
81. Mildred in Bred streete nere Chepside
82. Mildred in the Poultry
83. Nicholas Acons Nicholas lane nere Lüberstreet
84. Nicholas Cole Abby in old Fishstreet
85. Nicholas Olaves in Breadstreet
86. Olaues in Hart street nere Cruched friers
87. Olaues in old Iury at ye lower end of Chepside
88. Olaues in Silver streete
89. Pancras in Soper lane nere Bucklersbery
90. Peters nere Chepside
91. Peters in Cornehill
92. Peters nere Paules wharfe
93. Peters ye poore nere Brod streete
94. Steven in Coleman streete nere Moregate
95. Steven in Wallbrooke
96. Swithens in Cañon streete by London stone
97. Thomas ye Apostle
98. Trinitie Church aboue Quene Hith
99. Dutch Church nere Brodstreete

## OGILBY'S MAP OF LONDON

**Description.**—This is more exclusively a plan of the City than any we have yet considered. It runs roughly from the Tower to Lincoln's Inn Fields, and the reason why it is thus limited is that it was made as a survey to assist in the plotting out of land in the City after the Fire.

**Designer.**—John Ogilby was born about 1600, and did not turn his attention to surveying until he was about sixty-six, when he secured the appointment as "King's Cosmographer and Geographical Printer." He died in 1676, the year before his map was published. He was assisted in the work by William Morgan, his wife's grandson, and most of the actual engraving of the map was done by Hollar.

**Original.**—The original is 8 feet 5 inches by 4 feet 7 inches, and is in twenty sheets. It is on the scale of 100 feet to the inch. It may be seen in the British Museum (Crace Collection) and in the Guildhall. The two examples differ a little, and that in the Guildhall has an additional sheet. The reproduction here given is taken from that made by the London and Middlesex Archæological Society from the British Museum copy. The arms of the City are in the left-hand top corner, and those of Sir Thomas Davies, Lord Mayor 1676-77, in the right-hand corner.

**Details.**—Beginning at the left-hand top corner, we find pastures, bowling-greens, and market-gardens. Aylesbury House, next to St. John Street, has magnificent private gardens, and beyond the Charterhouse bowling-green there is a wood. Further east the Honourable Artillery Company, which had been revived by Cromwell, can be seen, with their equipment and tents. This company is directly descended from the Finsbury Archers, whom we noted in the last map, and it is interesting to know that the actual ground on which they are here depicted is still reserved for their use. Moorfields is neatly laid out and planned, and south of it is new Bethlehem Hospital, now transferred across the river. Eastward, again, there is a large open space at Devonshire House Garden, and southward innumerable gardens can be seen, some of which are preserved to this day behind City halls, etc., but so hidden that no one who did not know of their existence could possibly find them.

On tracing the line of the City wall on the north side we see how some of the churches, notably St. Giles's and St. Botolph's, have taken a part of the town ditch for the enlargement of their churchyards; near St. Bartholomew's the town ditch is still marked. This ditch caused the Mayor and Council as much worry as the increase of houses, because it was the receptacle for every kind of filth, and its cleansing annually swallowed up a large sum of money. The Fleet River is shown flowing down in the open, and is called the New Canal. It is crossed by a bridge at Holborn and another at Fleet Street. We can mark the sinuous line of the great thoroughfare of Holborn as it was before the viaduct and approaches were made. The Strand outside Temple Bar shows the obstructions which have only finally been removed in our own time. Butcher Row disappeared first in 1813; other streets followed to make way for the new Law Courts, and with the destruction of Holywell Row and the opening of Kingsway the improvements here may be considered complete.

To the south are the great houses of Essex and Arundel, with their gardens; their names are preserved in the streets that flow over their sites. Somerset House, the Protector's palace, was then standing, and did not make way for its present representative for another hundred years. The river is covered with wherries, clustered as thickly as ants. It is still the main highway for most people, though there were hackney coaches for hire. There was still only London Bridge by which to get across the river on foot, and the boats were used as ferries. There were tilt-boats, too, as well as the smaller wherries; these ran at stated intervals, like our own omnibuses, and were protected by an awning. Near the Fleet mouth is Bridewell, once a palace, and the scene of the meeting of Parliament, but given by Edward VI. to be a prison. On the east is a blank space, where is now the station of the London Chatham and Dover Railway Co., who purchased it in 1844. The site of St. Paul's was plotted out, but not yet built upon. In fact, the rebuilding of the houses was the first consideration, and was done with remarkable promptness, for in the meantime the poor houseless wretches were camping on Moorfields. The churches and city halls were therefore left to the last; yet even so we may see that, though only eleven years had elapsed since the destruction of the City, about twenty churches had been rebuilt out of the eighty-seven that were destroyed. The picturesque Old London of the gable-ends and overhanging stories was gone, never to return; but gone also was a great deal of rubbish and an insanitariness never afterwards quite so bad. As for the overcrowding, we must see what Sir Walter Besant says:

"If we look into Ogilby's map, we see plainly that as regards the streets and courts London after the Fire was very much the same as London before the Fire; there were the same narrow streets, the same crowded alleys, the same courts and yards. Take, for instance, the small area lying between Bread Street Hill on the west and Garlick Hill on the east, between Trinity Lane on the north and Thames Street on the south: is it possible to crowd more courts and alleys into this area? Can we believe that after the Fire London was relieved of its narrow courts with this map before us? Look at the closely-shut-in places marked on the maps—' 1 g., m. 46, m. 47, m. 48, m. 40.' These are respectively Jack Alley, Newman's Rents, Sugar-Loaf Court, Three Cranes Court, and Cowden's Rents. Some of these courts survive to this day. They were formed, as the demand for land grew, by running narrow lanes between the backs of houses and swallowing up the gardens. There were 479 such courts in Ogilby's London of 1677, 472 alleys, and 172 yards, besides 128 inns, each of which, with its open courts for the standing of vehicles and its galleries, stood retired from the street on a spot which had once been the fair garden of a citizen's house." (*London in the Time of the Stuarts*, p. 280).

THE FOLLOWING EXPLANATIONS ARE EXTRACTED FROM OGILBY'S KEY TO THE
MAP IN THE BRITISH MUSEUM

We Proceed to the Explanation of the Map, containing 25 Wards, 122 Parishes and Liberties, and therein 189 Streets, 153 Lanes, 522 Alleys, 458 Courts, and 210 Yards bearing Name.

The Broad Black Line is the City Wall. The Line of the Freedom is a Chain. The Division of the Wards, thus oooo. The Parishes, Liberties, and Precincts by a Prick-line, .... Each Ward and Parish is known by the Letters and Figures Distributed within their Bounds, which are placed in the Tables before their Names. ... The Wards by Capitals without Figures. The Parishes, &c., by Numbers without Letters. The Great Letters with Numbers refer to Halls, Great Buildings, and Inns. The Small Letters to Courts, Yards, and Alleys, every Letter being repeated 99 times, and sprinkled in the Space of 5 Inches, running through the Map, from the Left Hand to the Right, &c. Churches and Eminent Buildings are double Hatch'd, Streets, Lanes, Alleys, Courts, and Yards, are left White. Gardens, &c. faintly Prick'd. Where the Space admits the Name of the Place is in Words at length, but where there is not room, a Letter and Figure refers you to the Table in which the Streets are Alphabetically dispos'd, and in every Street the Churches and Halls, Places of Note, and Inns, with the Courts, Yards, and Alleys, are named; then the Lanes in that Street, and the Churches, &c. as aforesaid, in each Lane.

THE SEVERAL MARKS AND NAMES OF THE WARDS, PARISHES, AND LIBERTIES

WARDS

A Faringdon Without
B Faringdon Within
C Bainard-Castle
D Bread-Street
E Queen-Hith
F Cordwainers
G Walbrook
H Vintry

I Dowgate
K Broad-Street
L Cornhil
M Cheap
N Bassishaw
O Coleman-Street
P Bishopsgate
Q Cripplegate

R Aldersgate
S Billing-gate
T Lime-Street
U Langborn
W Portsoken
X Aldgate
Y Candlewick
Z Bridg

T Tower

PARISHES AND LIBERTIES

1. St. James Clerkenwel
2. St. Giles Cripple-Gate
3. St. Leonard Shoreditch.
4. Norton-Folgate Liberty
5. St. Botolph Bishopsgate
6. Stepney
7. St. Stephen Coleman Street
8. Alhallows on the Wall.
9. St. Andrew Holborn
10. St Giles in the Fields
11. St. Sepulchers
12. St. Mary Cole-Church

13. St. Botolph Aldersgate
14. St. Alphage
15. St. Alban Wood Street
16. St. Olave Silver Street
17. St. Michael Bassishaw
18. Christ Church
19. St. Anne Aldersgate
20. St. Mary Staining
21. St. Mary Aldermanbury
22. St. Olave Jewry
23. St. Martin Ironmonger Lane
24. St. Mildred Poultry

25. St. Bennet Sherehog
26. St. Pancras Soaper Lane
27. St. Laurence Jewry
28. St. Mary Magdalen Milk Street
29. Alhallows Hony Lane
30. St. Mary le Bow
31. St. Peter Cheap
32. St. Michael Wood Street
33. St. John Zachary
34. St. Martins Liberty
35. St. Leonard Foster Lane
36. St. Vedast, alias Foster

21

37. St. Michael Quern
38. St. John Evangelist
39. St. Matthew Friday Street
40. St. Margaret Lothbury
41. St. Bartholemew Exchange
42. St. Christophers
43. St. Mary Woolnoth
44. St. Mary Woolchurch
45. St. Michael Cornhil
46. St. Bennet Fink
47. St. Peter Poor
48. St. Peter Cornhil
49. St. Martin Outwich
50. St. Hellens
51. St. Ethelborough
52. St. Andrew Undershaft
53. Alhallows Lumbard Street
54. St. Edmond Lumbard Street
55. St. Dionis Back-Church
56. St. Katherine Cree-Church
57. St. James Dukes Place
58. St. Katherine Coleman
59. St. Olave Hart Street
60. St. Botolph Aldgate
61. St. Mary White Chapel
62. Trinity Minories
63. St. Bartholemew the Great.
64. Alhallows Staining
65. Alhallows Barking

66. St. Mary Abchurch
67. St. Nicholas Accorn
68. St. Clement East Cheap
69. St. Bennet Grace-Church
70. St. Gabriel Fenchurch
71. St. Margaret Pattons
72. St. Andrew Hubbart
73. Dutchy Liberty
74. St. Clement Danes
75. Rolls Liberty
76. St. Dunstan in the West
77. White Fryers Precinct
78. St. Bridget
79. Bridewel Precinct
80. St. Anne Black-Fryers
81. St. Martin's Ludgate
82. St. Gregories
83. St. Andrew Wardrobe
84. St. Bennet Paul's Wharf
85. St. Peter
86. St. Mary Magdaline Old Fish-Street
87. St. Nicholas Cole-Abby
88. St. Austine
89. St. Margaret Moses
90. Alhallows Bread-Street
91. St. Mildred Bread-Street
92. St. Nicholas Olave
93. St. Mary Mounthaw
94. St. Mary Somerset

95. St. Michael Queen Hith
96. Trinity
97. St. Mary Aldermary
98. St. Thomas Apostles
99. St. Michael Royal
100. St. James Garlick-Hith
101. St. Martin Vintry
102. St. Antholin's
103. St. John Baptist
104. St. Stephen Walbrook
105. St. Swithin
106. St. Mary Bothaw
107. Alhallows the Great
108. St. Faith's
109. St. Leonard East Cheap
110. St. Laurence Poultney
111. St. Martin Orgar's
112. Little Alhallows
113. St. Michael Crooked Lane
114. St. Magnus at the Bridg
115. St. Margaret New Fish-Street
116. St. George Botolph Lane
117. St. Botolph Billingsgate
118. St. Mary Hill
119. St. Dunstans in the East
120. Little St. Bartholemews
121. Tower Liberty
122. St. Katherines

## LIST OF PRINCIPAL BUILDINGS IN OGILBY & MORGAN'S MAP, 1677

### COMPILED FROM THE MAP AND KEY

The References on the left of the names refer to the marginal numbers on the Map

7-14. African House, Throgmorton Street, B55
2-5. Ailesbury's House, Earl of, A7
7-18. Alkgate
10-17. Alhallows Barking Church
9-10. Alhallows Bread-street Church
11-12. Alhallows Church, Great
11-12. Alhallows Church, Little
7-10. Alhallows Hony Lane Church [site absorbed into Hony Lane Market]
9-14. Alhallows Lombard Street Church
5-14. Alhallows on the Wall Church
9-17. Alhallows Staining Church, Mark Lane
9-6. Apothecary's Hall, C1
5-12. Armorers Hall, Coleman Street, A65
11-1. Arundel House

5-10. Barber Chyrurgeons Hall, A59
6-15. Barnadiston's House, Sir Samuel, B61
6-3. Barnard's Inn
6-3. Bell Inn, Holborn, A83
8-6. Bell Savage Inn, Ludgate Hill, B77
3-6. Berkley's House, Lord, A11
6-14. Bethlehem, New
0-15. Bishops Gate
6-3. Black Bull Inn, Holborn, A84
6-3. Black Swan Inn, Holborn, A81
1-10. Blacksmith's Hall, C29
7-10. Blewel Hall, B49
7-11. I'm Inn, B48
6-9. Howard's House, Sir Thomas, also Lord, B3

9-4. Bolt and Tun Inn, Fleet Street, B98
6-10. Brewers Hall, Addle Street, B7
8-17. Brick-Layers Hall, Leaden Hall Street, C52
9-6. Bridewell
9-6. Bridewel Precinct Chapel, Bride Lane
3-9. Bridgwaters House, Earl of, A18
6-2. Brook House
10-11. Buckingham's House, Duke of, C19
6-8. Bull and Mouth Inn, Bull and Mouth Street, A98
10-15. Butchers Hall, C39

9-2. Chancery Office, Chancery Lane, B73
3-6. Charter House
7-7. Christ Church, Newgate Street
7-7. Christ Hospital
7-12. Clayton's House, Sir Robert, Old Jewry, B52
9-1. Clements Inn
6-9. Clerks Hall, Silver Street, B4
9-3. Clifford's Inn
9-16. Cloth Workers Hall, Mincing Lane, C25
6-9. Cooks Hall, Aldersgate Street, C50
6-11. Coopers Hall, Bassishaw Street, B14
9-9. Cordwainers Hall
5-10. Cripple Gate
5-10. Curryers Hall, London Wall, A60
7-2. Cursitor's Office

11-17. Custome house
9-12. Cutlers Hall, Cloak Lane, C21

6-5. David's House, Sir Thomas, Snow Hill, B34
5-16. Devonshire House, A73
9-9. Doctors Commons, C10
3-7. Dorchester's House, Marquess of, A13
7-14. Drapers Hall, B57
6-14. Dutch Church
11-13. Dyers Hall, New Key, Thames Street

8-16. East India House, Leaden Hall Street, B88
6-4. Ely House
10-1. Essex House
6-14. Excise Office, Broad Street, C60

10-15. Fiery Pillar, The [The Monument]
11-14. Fishmongers Hall, Thames Street
9-6. Fleet Bridg
8-5. Fleet [Prison]
7-12. Founders Hall, Loathbury, B56
7-12. Frederick's House, Sir John, Old Jewry, B51
7-14. French Church, B62
6-3. Furnival's Inn

6-6. George Inn, Holborn Bridg, A92
9-10. Gerrard's Hall Inn, C16
5-11. Girdlers Hall, A63
3-10. Glovers Hall, Beech Lane, A20

7-9. Goldsmiths Hall, Foster Lane, B39
5-1. Gray's Inn
7-15. Gresham Colledge
3-7. Grey's House, Lord, A14
8-12. Grocers Hall, B53
7-11. Guild Hall

7-10. Haberdashers Hall, B8
7-12. Hern's House, Sir Nathiel, Loathbury, B54
4-6. Hicks's Hall
7-5. Holborn Bridge
— [Holy] Trinity Church, Trinity Lane [see Trinity Church]
— [Holy] Trinity Minories Church [see Trinity Minories]

9-3. Inner Temple, Inner Temple Lane
10-12. Inn-Holders Hall, Elbow Lane, C34
8-17. Ironmongers Hall, Fenchurch Street, B91
11-11. Joyners Hall, Fryer Lane, Thames Street, C37

6-5. Kings Arms Inn, Holborn Bridg, A90
9-7. King's Printing House, C3

5-11. Lariner's Hall, Fore Street, A78
7-16. Lawrence's House, Sir John, Great St. Hellens, B67
8-15. Leaden Hall Market
6-16. Leather-Sellers Hall
7-2. Lincoln's Inn
10-1. Lions Inne
11-14. London Bridg
5-8. London House, A57
9-7. Ludgate
9-10. Lutheran Church, Trinity Lane (N.E. corner Little Trinity Lane)

8-11. Mercer's Chapel
8-14. Merchant-Taylors Hall
10-12. Merchant-Taylors School, Suffolk Lane, C39
9-3. Middle Temple, Middle Temple Lane
8-10. Milkstreet or Hony lane Market
— [Monument, The, see "Fiery Pillar"]

9-17. Navy Office, Mark Lane, C26
10-1. New Inn
2-4. New Prison, or Bridewel, Clerkenwel Green
2-4. Newcastle's House, Duke of, A6
7-6. Newgate
8-7. Newgate Market

10-10. Painters Stainers Hall
8-17. Papillion's House, Mr. Tho., Fenchurch Street, C54
6-14. Pay Office, Broad Street, B22
8-16. Pewterers Hall, Lime Street, C62
7-7. Physicians Colledge, B37
6-14. Pinner's Hall, B21
6-10. Plaisterers Hall, Addle Street, B6

6-15. Post Office, General, Bishopsgate Street Within, B59
8-12. Poultry Compter, B83
9-8. Prerogative Office, St. Paul's Church Yard, C6

8-4. Red Lyon Inn, Fleet Street, B75
7-5. Rose Inn, Holborn-Bridg, A91
8-14. Royal Exchange

7-9. Sadler's Hall, Cheapside, B41
9-13. Salter's Hall, St. Swithins Lane, C23
6-5. Sarazens Head Inn, Snow Hill, A93
9-6. Scotch Hall, C2
6-9. Scriveners Hall
9-3. Serjeant's Inn, Chancery Lane, B07
9-4. Serjeant's Inn, Fleet Street
8-6. Session House, The, Old Bayly
9-8. Sheldon's House, Sir Joseph, St. Paul's Church Yard, C7
8-2. Simond's Inn, Chancery Lane, B71
5-11. Sion College, A61
9-2. Six Clarks Office, Chancery Lane, B72
10-12. Skinners Hall, Dough-Gate Hill, C33
5-6. Smithfield Penns
11-1. Somerset House
6-10. St. Alban Wood-Street Church
5-11. St. Alphage Church, LondonWall
6-4. St. Andrew Holborn Church
10-15. St. Andrew Hubbart Church, Little East-Cheap [formerly S. side, between Buttolph Lane and Love Lane]
8-16. St. Andrew Undershaft Church, Leaden Hall Street, B66
10-7. St. Andrew Wardrobe Church
6-9. St. Anne Aldersgate Church
9-6. St. Anne Black-Fryers Church
9-12. St. Antholine's Church, Budg Row
8-9. St Austine's Church
5-7. St. Bartholemew Church, Great
6-7. St. Bartholemew's Church, Little
8-13. St. Bartholemew Exchange Church
6-7. St. Bartholemew's Hospital
8-13. St. Bennet Fink Church
8-15. St. Bennet Grace Church
10-8. St. Bennet Pauls Wharf Church
8-11. St. Bennet Sherehog Church
9-6. St. Bridget's Church
6-9. St. Buttolph Aldersgate Church
6-19. St. Buttolph Aldgate Church
11-15. St. Buttolph Billingsgate Church [formerly S. side of Thames Street between Buttolph Lane and Love Lane]
5-11. St. Buttolph Bishopsgate Church
8-13. St. Christophers Church
10-1. St. Clement Danes Church
9-14. St. Clement's Eastcheap Church
9-3. St. Dunstan's Church
10-16. St. Dunstan's in the East Church
9-14. St. Edmond Lumbard Street Church

6-16. St. Ethelborough Church, Bishopsgate Street Within [immediately N. of Little St. Hellens]
9-8. St. Faith's Church [under-St.-Paul's]
9-16. St. Gabriel Fenchurch Church [absorbed into the roadway of Fenchurch Street, between Rood Lane and Mincing Lane]
10-15. St. George Buttolph Church, C40
4-10. St. Giles's Cripplegate Church
9-8. St. Gregory's Church [site absorbed by St. Paul's]
7-16. St. Hellen's Church
7-18. St. James Dukes Place Church, Dukes Place
10-11. St. James Garlick Hith Church
9-12. St. John Baptist Church
9-9. St. John Evangelist Chnrch, Friday Street [formerly E. side, at the corner of Watling Street, having the latter street on the north]
6-9. St. John Zachary Church, Maiden Lane
8-17. St. Katherine Coleman Church
8-17. St. Katherine Cree Church, Leaden Hall Street, B68
10-13. St. Laurence Poultney Church
7-11. St. Lawrence Jewry Church
10-15. St. Leonard East Cheap Church
7-9. St. Leonard Foster-Lane Church
11-14. St. Magnus Church, Thames Street, C59
9-13. St. Mary Abchurch Church
6-11. St. Mary Aldermanbury Church
9-11. St. Mary Aldermary Church
9-12. St. Mary Bothaw Church
6-11. St. Mary Cole Church, Cheapside [formerly S.W. corner of Old Jewry]
10-16. St. Mary Hill Church C43
8-10. St. Mary le Bow Church
7-10. St. Mary Magdalen's Church, Milk Street [site absorbed into Hony lane Market]
10-9. St. Mary Magdaline Old Fish Street Church
10-9. St. Mary Mountshaw Church
10-9. St. Mary Somerset Church
6-9. St. Mary Staining Church, Oat Lane
8-12. St. Mary Wool Church [site absorbed into Wool Church Market]
8-13. St. Mary Woolnoth Church, Lumbard Street [opposite Pope's Head Alley]
7-12. St. Margaret Loathbury Church
9-9. St. Margaret Moses Church, Friday Street [formerly S.W. corner of Basing Lane]
9-15. St. Margaret Patton's Church
10-15. St. Margaret's New Fish Street Church [site absorbed by the Monument]
7-11. St. Martin Ironmonger Church, Ironmonger Lane [formerly adjoining the west end of St. Olave Jewry]
8-7. St. Martin Ludgate Church

10-13. St. Martin Or r's Church
7-15. St. Martin Outwich Church, Bishopsgate Street Within [S.E. corner of Thread Needle Street]
10-11. St. Martin Vintry Church
8-9. St. Mathew Friday Street Church
9-10. St. Mildred Bread-Street Church
8-12. St. Mildred Poultry Church, B84
. 11. St. Michael Pater Noster Church
8-14. St. Michael Cornhill
10-14. St. Michael Crooked Lane Church
10-10. St. Michael Queen Hith Church
7-9. St. Michael Quern Church, Cheapside [site absorbed into roadway of Cheapside at junction of Pater Noster Row and Blow Bladder Street]
9-11. St. Michael Royal Church
7-9. St. Michael Wood-Street Church, B45
9-13. St. Nicholas Acorn Church
9-9. St. Nicholas Cole-Abby Church, Old Fish Street (N.W. corner of Old Fish St. Hill)
9-10. St. Nicholas Olave's Church, Bread-Street Hill [formerly near middle of W. side]
9-17. St. Olave Hart-street Church, C27
7-12. St. Olave Jewry Church

5-10. St. Olave Silver Street Church
8-11. St. Pancras Soaper Lane Church
9-8. St. Paul's Cathedral
9-8. St. Paul's House, Dean of, St. Paul's Church Yard, C5
11-18. [St. Peter-ad-Vincula] Church, Tower of London
7-10. St. Peter Cheap Church
6-14. St. Peter Poor Church
10-8. St. Peter's Church
8-14. St. Peter's Cornhil
7-6. St. Sephlcher's Church
6-12. St. Stephen Coleman Street Church, B56
9-12. St. Stephen Walbrook Church
10-12. St. Swithin Church, Cannon Street
9-11. St. Thomas Apostles Church, St. Thomas Apostles
7-9. St. Vedast Church, B40
6-2. Staple Inn
8-7. Stationers Hall
6-5. Swan Inn, Holborn-Bridg, A89
6-10. Swan with Two Necks Inn, Ladd Lane, B11
9-12. Tallow Chandlers Hall, Dough-Gate Hill, C22
10-3. Temple Church
5-9. Thanet House, A58
6-4. Thavy's Inn, Holborn, A86

11-19. Tower, The
— Trinity Church, Trinity Lane [site occupied by Lutheran Church, which see]
10-17. Trinity House, Water Lane, C45
8-19. Trinity Minories Church, B70
9-8. Turners House, Sir William, St. Paul's Church Yard, C4
11-11. Vintonners Hall
8-13. Vyner's House, Sir Robert, Lumbard Street, B85
10-13. Ward's House, Sir Patient, Lawrence Poultney's Hill, C38
6-1. Warwick House
11-13. Watermans Hall, New Key, Thames Street, C28
11-13. Waterman's House, Sir George, Thames Street, C57
7-10. Wax Chandellors Hall, Maiden Lane, B43
6-11. Weavers Hall, Basishaw Street, B13
8-17. Whitchurch House, Leaden Hall Street, C53
10-11. Whittington's College, College Hill, *m*15
7-10. Wood Street Compter, B46
9-12. Wool Church Market

# LONDON IN 1741-45

## By JOHN ROCQUE

**Description.**—In some ways this map is the most interesting of the whole series, for it comes nearest to our own times, and yet by studying it we can infer the remarkable changes that have taken place within the memory of man. It is much more comprehensive than Ogilby's, including the whole of the outlying suburbs, and even going as far as Edgware and Tottenham, which are still no part even of Greater London.

**Designer.**—Very little is known about John Rocque. He was probably a native of France, but was residing in England about 1750. He engraved maps and a few views from his own designs.

**Original.**—The original is in twenty-four sheets, and is 13 feet in length and $6\frac{3}{4}$ feet in depth. It can be seen at the British Museum. That which is here presented is the central part of this, not reduced, but on the same scale. Its interest is greatly increased by the fact that the names are printed on the map, and are not given separately as in other instances. To facilitate this Rocque has marked the houses bordering streets in white, and only blocked them in black where they line market-gardens and other parts indicated by a light surface. The map is a model of care and comprehensive detail.

**Detail.**—Beginning in the lower left-hand corner, we have the Royal Hospital, with its neatly-laid-out grounds. Close to it the Westbourne, whose irregular line determined the boundaries of Chelsea, falls into the Thames; higher up its course is through the Five Fields, now one of the most wealthy and popular districts of London —namely, Belgravia. St. George's Hospital is already standing at Hyde Park Corner, and a fringe of houses lines the road to Knightsbridge. Westminster is still largely open in the west by Tothill Fields, scene of so many tournaments and jousts, and the curve of the river encloses innumerable market-gardens. In St. James's Park the stiff canal, memento of Dutch influence, has not yet been transformed into the more attractive ornamental water. Carlton House Terrace has not come into existence. Here Carlton House, which does not appear to be marked, was standing, and was occupied by Frederick, Prince of Wales, father of George III. North of this, with the omission of Regent Street, made in 1813-20, the streets are pretty much as we know them. It is beyond Oxford Street northward that the difference is striking. This district was only just being built upon, and the well-laid-out streets soon run off into open country. "Marybone" Gardens, a favourite tea-garden, and the church, and a few houses, form a little hamlet just connected with the other part of London by a single street, and further westward, north of Berkeley Square, are fields. In the midst of these is the "Yorkshire Stingo," the public-house from which the first omnibus in the Metropolis began to run in 1829. The Tyburn Gallows still had much work to do; it was fifty years later that the last execution took place here. Just within the Hyde Park is the gruesome record, "where soldiers are shot." If we follow Oxford Street eastward to Tottenham Court Road, we find that it is only connected with High Holborn by the curve through High and Broad Streets at St. Giles's. To the south is the star of Seven Dials, and all the district so completely altered by the cutting through of Charing Cross Road, and then Shaftesbury Avenue in modern times. To the north, Montagu House occupies the site the British Museum was destined to fill; it was purchased by the Government in 1753, and pulled down about a hundred years later. Bedford House, the town residence of the Dukes of Bedford, stood until 1800. Behind, Lamb's Conduit Fields run up to Battle Bridge, where one of the early British battles was fought; this is now the site of King's Cross Station. Not far off Bagnigge Wells and Sadler's Wells are in the heyday of their prosperity. The Fleet or River of Wells may be traced passing through the former, but further south it is covered in, and does not appear in the open again until below Fleet Bridge, when it is ignominiously called Fleet Ditch.

Thames side is still fringed with "stairs to take water at" leading from the great houses on the margin, and there is as yet no embankment. Westminster and Blackfriars Bridges, however, afford easy access to the southern side. The labyrinth of the City is not seriously different from that of the present day except in the omission of Cannon Street. Bethlehem Hospital is still conspicuous, and the City wall has vanished strangely. What we now call Finsbury Square is marked as Upper Moorfields. We have to go far before we clear the houses to the east. Stepney and Bethnal Green are fairly thickly populated, and though surrounded by open ground, are connected by houses all the way from the City. But in the bend of the river by Wapping the chief area is occupied by market-gardens. Crossing over to the other side, we find the market-gardens very prominent; as London grows larger she thrusts her sources of supply further from her. The central ganglion of the Borough Road and its ray-like connections are marked out. At one end is the "King's Bench," which was close to the Marshalsea, associated with "Little Dorrit." The Marshalsea itself is not marked. Dickens was yet to come, and it was only through his writings that it gained a sentimental interest. A great part of the Borough is very marshy indeed, and we note frequent ponds. The "Dog and Duck," otherwise "St. George's Spaw," is almost surrounded by them.

To sum up in Sir Walter Besant's words :

"London, then, in the eighteenth century consisted first of the City, nearly the whole of which had been rebuilt after the Fire, only a small portion in the east and north containing the older buildings ; a workmen's quarter at Whitechapel ; a lawyer's quarter from Gray's Inn to the Temple, both inclusive ; a quarter north of the Strand occupied by coffee-houses, taverns, theatres, a great market, and the people belonging to these places ; an aristocratic quarter lying east of Hyde Park ; and Westminster, with its Houses of Parliament, its Abbey, and the worst slums in the whole City. On the other side of the river, between London Bridge and St. George's, was a busy High Street with streets to right and left ; the river bank was lined with houses from Paris Gardens to Rotherhithe ; there were streets at the back of St. Thomas's and Guy's ; Lambeth Marsh lay in open fields, and gardens intersected by sluggish streams and ditches ; and Rotherhithe Marsh lay equally open in meadows and gardens, with ponds and ditches in the east. . . .

"From any part of London it was possible to get into the country in a quarter of an hour. One realizes the rural surroundings of the City by considering that north of Gray's Inn was open country with fields ; that Queen Square, Bloomsbury, had its north side left purposely open in order that the residents might enjoy the view of the Highgate and Hampstead Hills. On the south side of the river Camberwell was a leafy grove ; Herne Hill was a park set with stately trees ; Denmark Hill was a wooded wild ; the hanging woods of Penge and Norwood were as lovely as those that one can now see at Cliveden or on the banks of the Wye" (*London in the Eighteenth Century*, pp. 77-79).

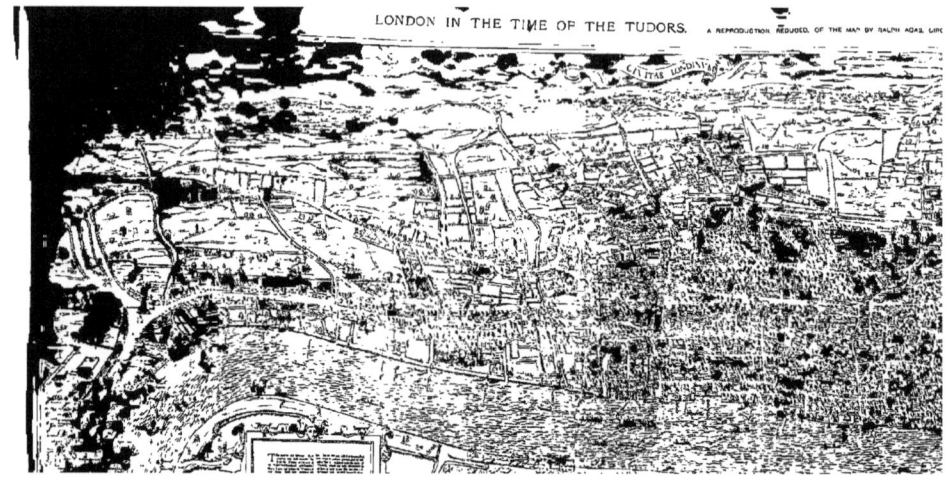
LONDON IN THE TIME OF THE TUDORS. A REPRODUCTION, REDUCED, OF THE MAP BY RALPH AGAS, 1560.

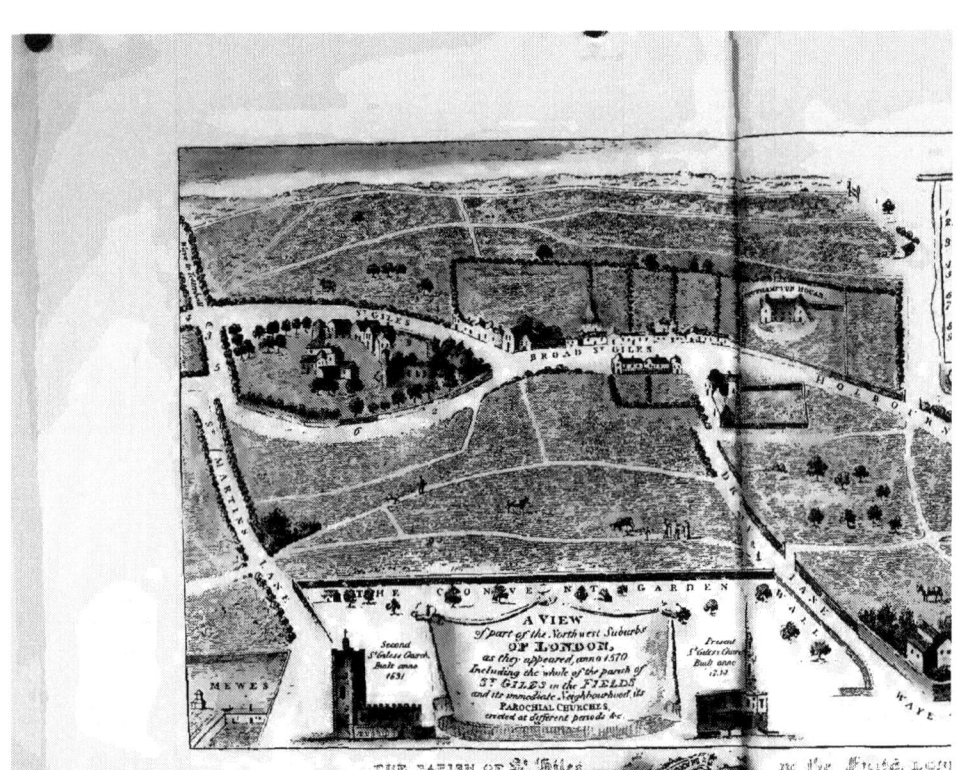

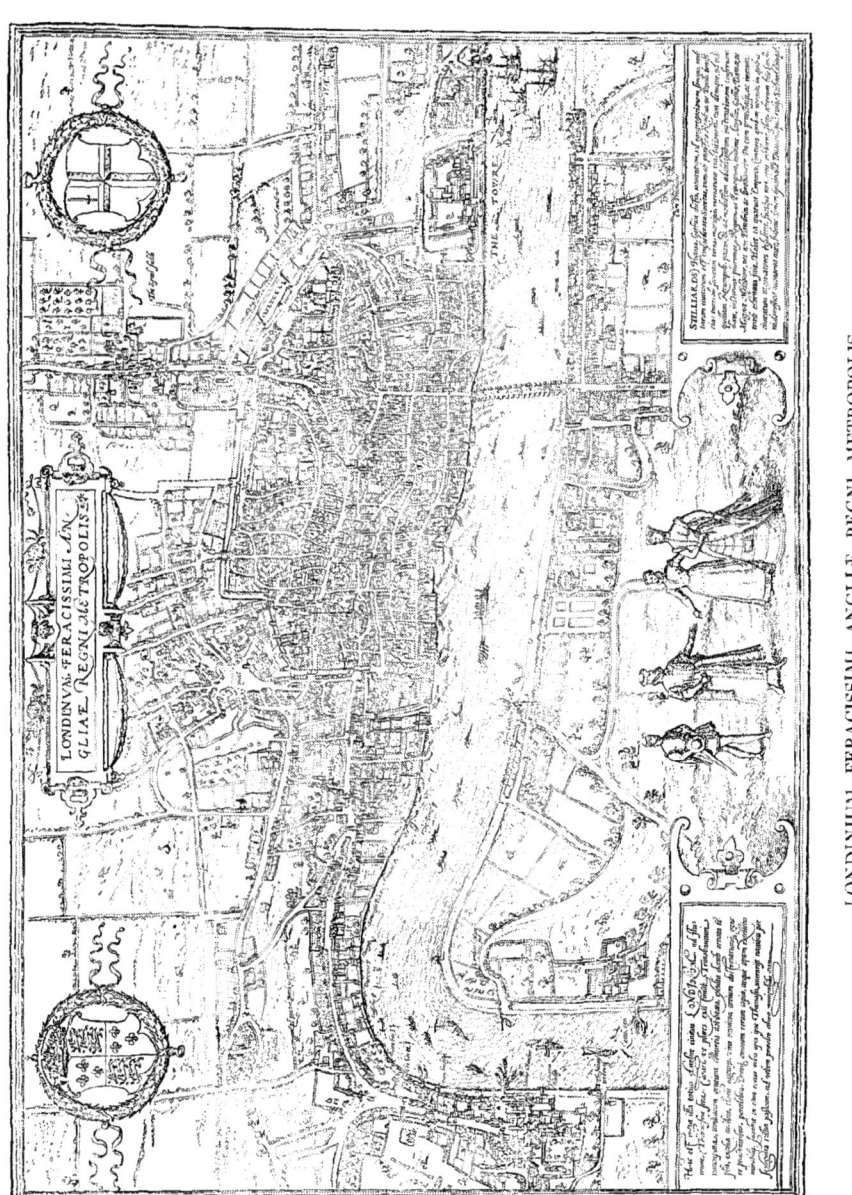

LONDINIUM FERACISSIMI ANGLIÆ REGNI METROPOLIS.

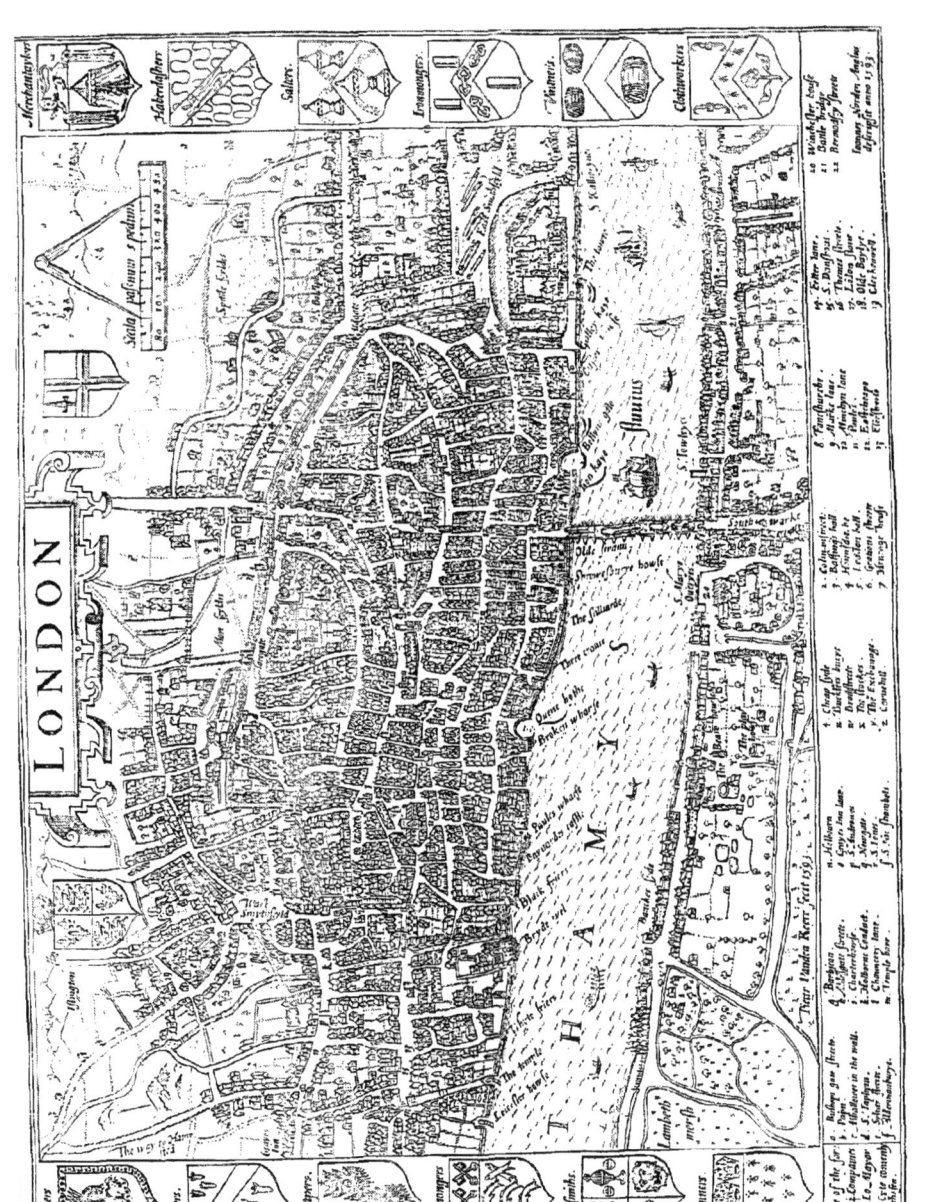

LONDON, 1593. BY JOHN NORDEN.

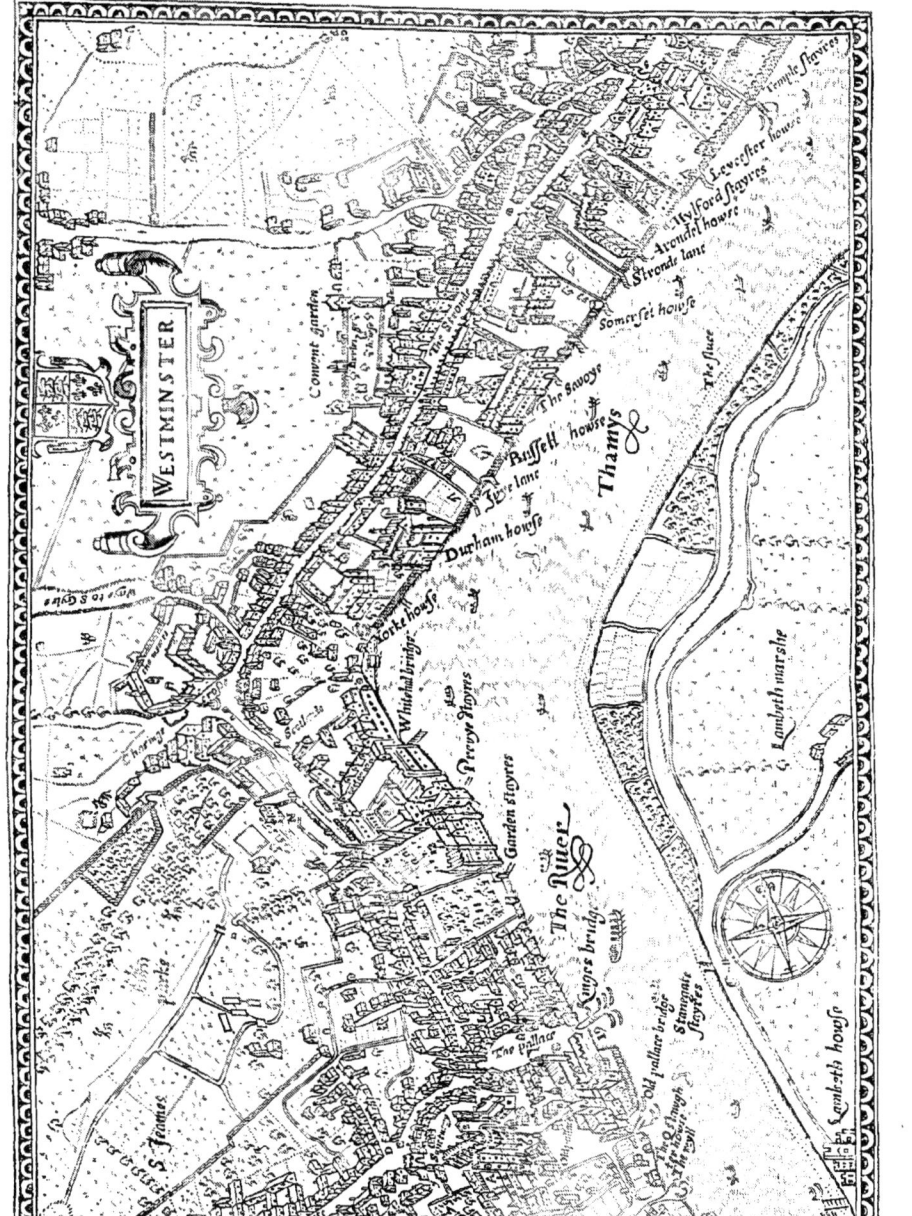

WESTMINSTER, 1593. By JOHN NORDEN.

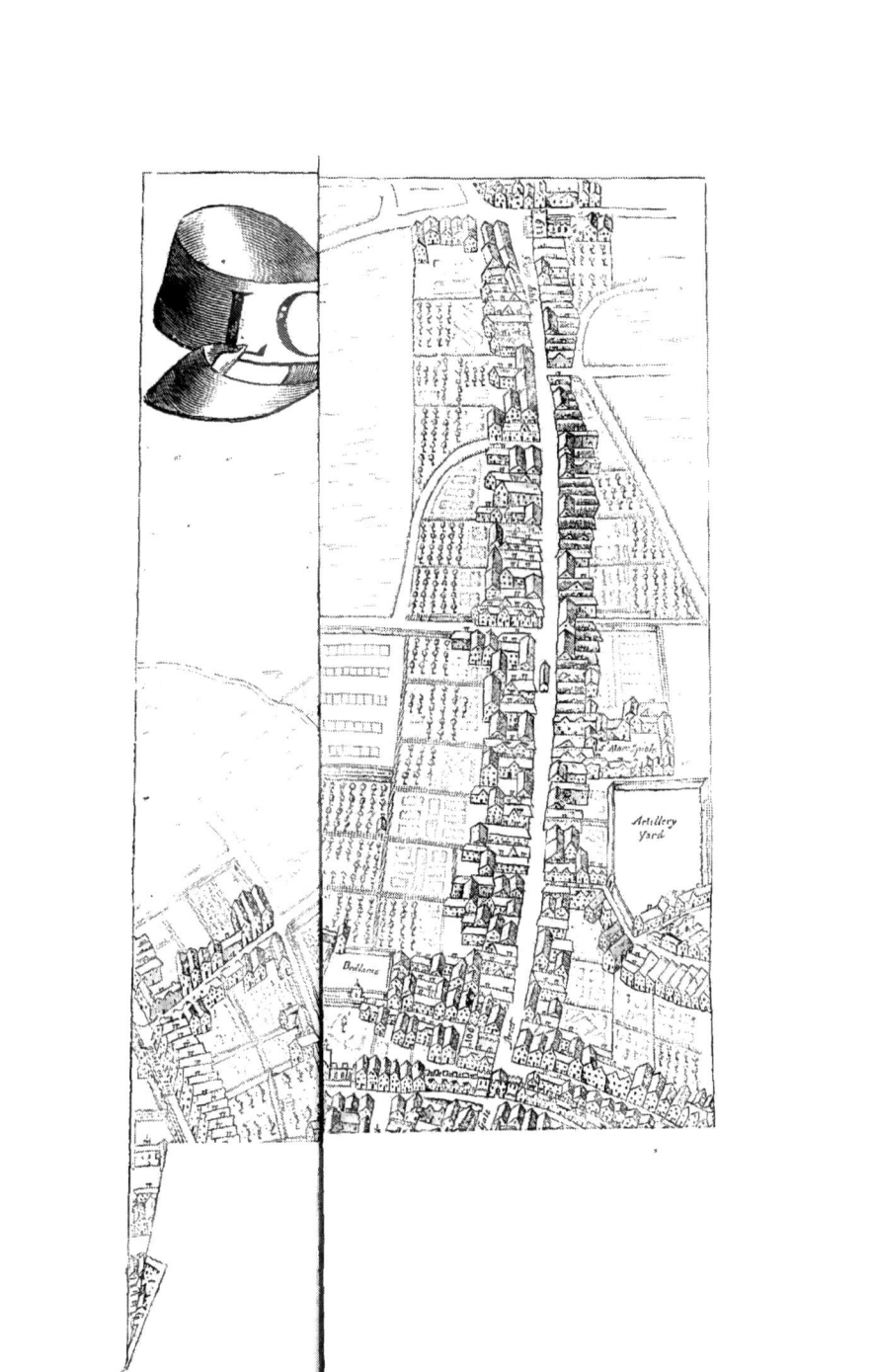

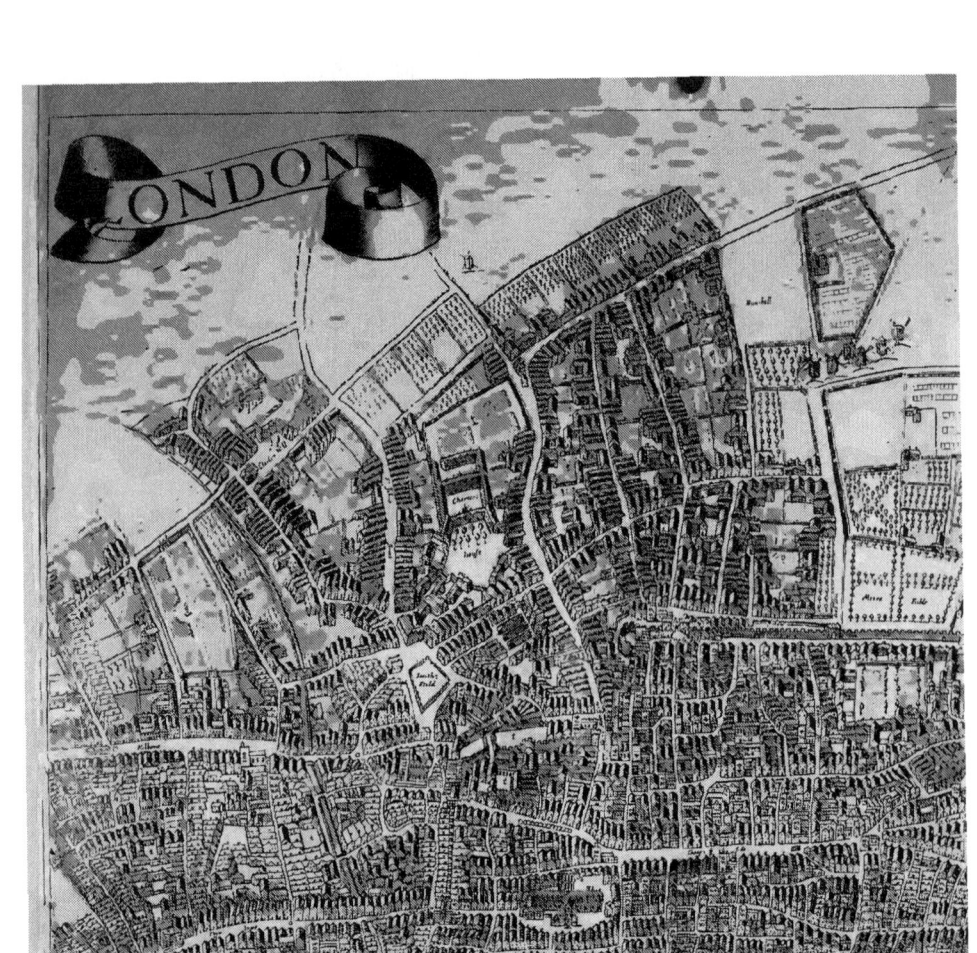

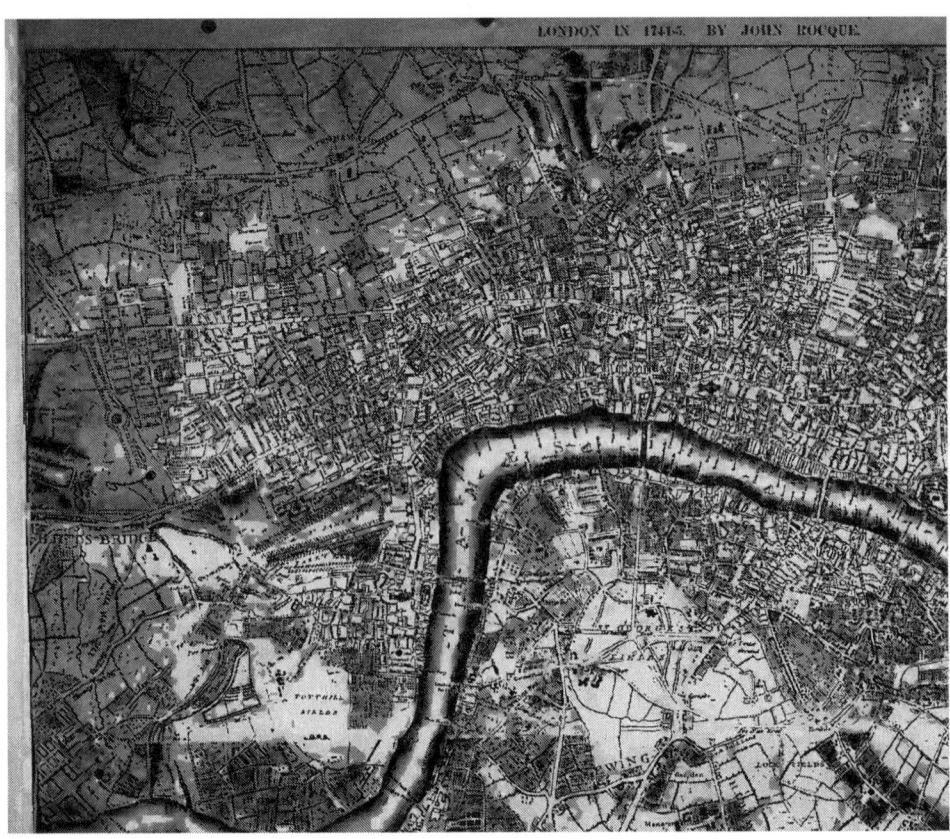

Milton Keynes UK
Ingram Content Group UK Ltd.
UKHW051841070823
426310UK00033B/167